DON'T FORGET THE ENVIRONMENT

A Guide for incorporating Environmental Assessment into your Project

Published by Institution of Chemical Engineers for The Secretariat of the Environmental Analysis Co-operative.

Applications for reproduction should be made to Institution of Chemical Engineers, Davis Building, 165-189 Railway Terrace, Rugby, CV21 3HQ, UK.

ISBN 0 85295 422 0

Representatives of the following companies, Government departments and other organisations contributed to the work of the Co-operative and to the publication of this document.

AEA Technology Plc
Assn. of British Pharmaceuticals Industries
BNFL
BOC Group
BP Chemicals
British Gas Transco
British Steel Strip Products
Cheshire County Council
Ciba Speciality Chemicals Plc
Contract Chemicals
Department of Trade and Industry
DoE Welsh Office
Environment Agency
Environment and Heritage Service (DoE)
Environmental Services Association
Federation of Small Businesses
Fluor Daniel Ltd
Hewlett-Packard Limited
Hill Consultants Ltd
ICI C&P
ICI Group
IMI Refiners
Institute of Environmental Assessment
Institute of Terrestrial Ecology
Jacobs Engineering
Jaguar Cars Ltd
Kent County Council
Ministry of Agriculture, Fisheries and Food
National Power Plc
National Society for Clean Air (NSCA)
NETCEN
Non-Ferrous Alliance
Ove Arup and Partners
Phillips Petroleum Company UK Ltd
Pilkington Glass Ltd
Powergen
Rechem International
Rhône-Poulenc Ltd
RIVM
Royal Town Planning Institute
Scottish Environmental Protection Agency
SGS Environment
Soil Survey and Land Research Centre
Solutia UK Ltd
Thames Water Utilities Ltd
UK Steel Association (British Steel Gas)
Water Services Association
Welsh Office
WRc Plc
WS Atkins Environment
Zeneca

This guide is based on the report of a Co-operative Work Group on interfaces between the planning process and environmental protection legislation. The Work Group was chaired by Rodney Perriman and conducted its work between March 1997 and March 1998.

Printed in the United Kingdom by Hobbs the Printers Limited, Totton, Hampshire.

CONTENTS

Page

Preface

This book is the second publication from the Environmental Analysis Co-operative (EAC). The EAC is an informal organisation of more than fifty industrial companies and trade associations, consultants, departments of central and local government, the Environment Agencies and the National Society for Clean Air. It came together in February 1995 with the aim of producing guidance for applicants seeking authorisation of processes under Integrated Pollution Control (IPC).

The first phase of its work was developing guidance for compiling an inventory of releases from a process and estimating the resulting levels of released substances in the local environment. That guidance was published by HMSO (now The Stationary Office) in March 1996 under the title "Released Substances and their Dispersion in the Environment" (RSDE) (Ref. 1).

Since RSDE was published there has been continuing debate about techniques for assessing the environmental impacts of releases from industrial processes and how to compare the relative overall environmental impacts of alternative processes. Over the same period, developments in environmental legislation and environmental management systems have highlighted the need for companies to incorporate environmental considerations into all aspects of business management. There has been increasing awareness of the linkages between pollution control and Town & Country Planning and the need for more effective co-ordination between pollution regulation and land use planning.

An EAC working group was set up in March 1997 to examine the implications of these issues for project development in manufacturing industries. The group decided to approach the subject by considering what advice would be most useful to help project managers, in companies without in-house specialist environmental support, to incorporate environmental considerations in their internal project development procedures and programmes. This book is based on the working group's report to the EAC.

The principles and techniques we recommend are based primarily on the requirements of IPC, under the Environmental Protection Act, 1990. We believe they are also relevant for IPPC, which comes into force in the UK in October 1999, and for preparing environmental information in support of planning applications under the Town & Country Planning Acts.

This guidance is published independently of the Environment Agency and the DETR, although representatives from the Agency and from planning authorities have been involved in its production as members of the Co-operative. The techniques and approaches recommended will not guarantee a successful IPC/IPPC or Planning application; nor does the guidance over-ride any legal requirements or official guidance from the Agency or government.

Throughout this book all references to English legislation and administrative arrangements and to the Environment Agency (for England & Wales) apply equally to corresponding legislation and regulatory agencies in Scotland and Northern Ireland.

References

1. Released Substances and their Dispersion in the Environment, published by HMSO, ISBN 0 11 702010.

The aims of the Environmental Analysis Co-operative are:

- To develop practical guidance to help operators make effective assessments of the environmental effects of processes;
- To improve the quality of environmental assessments in support of regulatory permits;
- To encourage good environmental management practices.

In keeping with these aims and its ethos of pursuing them through consensus and pooling experience, the EAC welcomes comments on its publications. If you would like to comment on any aspect of this guide, please write to:

Publications
IChemE
165-189 Railway Terrace
Rugby
CV21 3HQ
UK

1. INTRODUCTION

All new industrial developments and major changes to existing facilities normally require planning consent under the Town & Country Planning Act 1990 before construction can start.
Many industrial processes also require authorisation under the Integrated Pollution Control (IPC) requirements of the Environmental Protection Act 1990 before they can start up. From October 1999, these processes and many others will require authorisation under new UK legislation to implement the EC Directive (96/61/EC) on Integrated Pollution Prevention and Control (IPPC). Existing processes subject to IPC will be transferred to IPPC control by 2007, so both regimes will operate in parallel for some time.

At the present time (December 1998) UK regulations and official guidance for implementing the IPPC Directive are not available. However, we believe the recommendations in this guide are relevant for both the present requirements of IPC and the future requirements of IPPC. We use the acronym "IPC/IPPC" to refer to both pollution control regimes.

The procedures for obtaining both types of permits - planning consent and pollution control authorisation - involve assessment, by the relevant authorities, of the environmental implications of the proposed development at the chosen site. The purpose and scope of each type of permit is different.

The primary purpose of the land use planning system is to answer the question, *"is this an appropriate location for a development of this type?"*

The primary purpose of IPC/IPPC is to answer the question, *"given a development of this type in this location, what controls must be incorporated to achieve a high level of protection of the environment?"*

From these very basic statements it is clear that there are both overlaps and differences between the environmental aspects which each considers. Figure 1 illustrates some of the main areas of overlap and difference.

Figure 1 Main environmental considerations in planning and IPC/IPPC approval procedures

"Environmental Considerations"	IPC/IPPC	Planning
Land use		****
Visual impact	*	***
Raw materials	**	*
Energy	***	*
Emissions	****	***
Wastes	****	**
Accidental releases	**	*
Technology/controls	****	
Noise	**	****
Traffic		****
Economic & social aspects		***

The number of stars illustrates the typical relative importance of each aspect between the two regulatory systems.

The key to a successful application for planning consent or for IPC/IPPC authorisation is to demonstrate that all the environmental aspects of a proposed development have been considered. You must provide thorough and well presented assessments to justify the location and design of the proposed facility in relation to other practicable options that you might have chosen. You can only do this if you have taken account of all the environmental aspects of your project from the earliest stages of your development programme.

The challenge of any new industrial development is to achieve the best outcome for the environment as a whole, which is technically and commercially viable at the proposed site.

This guide provides a structured approach for making the evaluations, comparisons, informed judgements and assessments that should be an integral part of your organisation's project development programme in order to achieve this objective.

This guide is intended for managers in industry who are likely to require planning consents and IPC/IPPC authorisations for new developments or changes to existing installations. It will help you to:

- identify environmental aspects during the development of a project;
- incorporate environmental considerations as essential factors at all stages of your project development programme;
- assess the relative environmental effects of alternatives for process design and site selection;
- prepare environmental assessments in support of planning and/or IPC/IPPC applications.

In this guide the term "environmental aspects" covers not only the pollution control aspects of industrial installations, covered by the IPC/IPPC legislation, but also the wider environmental impacts which are often key issues for planning consent. We offer advice on ways to identify and manage these aspects and we recommend informal consultation with officials and local community interests at appropriate stages in your project development programme.

These recommendations should help you avoid problems and delays in progressing your applications for planning consent and IPC/IPPC authorisation. They will not, of course, guarantee that your applications will always be successful but the systematic approach, with good record keeping, should help you to clarify and resolve any differences of views with the authorities and other interested parties.

The scope of this guide is assessment of the environmental effects of a project operating under normal, or design, conditions. It does not address, in any detail, the assessment of risks to people or the environment from accidental events.

This is an important aspect of environmental assessment for many types of industrial development. It ranges from statutory requirements for formal risk assessments at installations classified as Major Accident Hazards, to internal company procedures for identifying the risks of incidents that could cause environmental harm and making sure that suitable control measures are in place to minimise those risks. Detailed advice on this aspect of environmental management is not included in this guide but useful references are given in Annex A.

2. HOW TO USE THIS GUIDE

Before you use this guide for an actual development we recommend that you read all the sections to understand how the procedures for environmental assessment can be incorporated into your organisation's project management process.

The relationship between land use planning and pollution control

Applications for planning consent are often delayed, or even refused, because the developer has not considered all the implications and concerns that the project might cause in the immediate neighbourhood. If a developer has not identified these issues and shown that they have been considered in his environmental assessment of the project, delay is inevitable at the planning application stage. Project managers in industry tend to focus on the technical and financial aspects of pollution control and overlook other aspects of the project that can turn out to be major stumbling blocks during public consultation on the planning application. By that stage of a project, delays and changes are likely to be very costly.

To help you appreciate this aspect, Chapter 3 of this guide - Planning and Pollution Control - provides a summary of the overlaps and distinctions between the two regimes. Annex A gives references to other publications on this subject that may also be helpful.

Chapter 3: Planning and pollution control

3.1 Planning consent
The purpose and scope of land use planning in relation to industrial development.

3.2 IPC/IPPC authorisation
The purpose and scope of IPC/IPPC authorisations.

3.3 Linkages, overlaps and distinctions between Planning and IPC/IPPC
How the two permitting systems are inter-related; the roles of the relevant authorities; the implications for industrial developments.

Incorporating environmental assessment into project development

To repeat the point made in the Introduction, the key to successfully managing the environmental aspects of a new development is to ensure that all relevant environmental aspects are considered at every stage of your project development programme, from the initial feasibility study through to the final engineering design. If this is not done there is likely to be costly delay because the authorities will not be able to reach a decision on the limited information you have provided.

Figure 2A shows a typical project development programme, from initial concept through design, construction and commissioning to operation and eventual closure. The right hand side of the diagram shows when formal applications for planning and pollution control consents have normally been initiated. All too often in the past, this was the first time in the project programme that serious consideration was given to the environmental effects of the project.

Figure 2B shows the project development programme with the environmentally related procedures that we recommend should be an integral part of project development practice.

Guidance for carrying out these procedures is provided in Chapter 4. This is the main part of this book. It provides a structured approach for developing a project with confidence that it can be justified to the regulating authorities and the local community as the best overall outcome for the environment, which is technically and commercially viable at the proposed site.

Chapter 4: Environmental assessment in project development

4.1 Introduction

4.2 Preliminary environmental review
A wide ranging but structured review to identify all the environmental issues that might affect the viability of the project, or are likely to be significant aspects of land use planning and/or IPC/IPPC.

4.3 Assessment of site and process options
A procedure for identifying and comparing the environmental effects of alternatives for meeting the project objectives. Techniques are described for determining which option has the least overall environmental impact.

4.4 Costs and environmental benefits
Advice for deciding whether project costs for reducing environmental impact are excessive.

4.5 Compiling and presenting environmental aspects: Planning applications, IPC/IPPC applications
Advice on compiling and presenting the relevant parts of the environmental information developed in the previous internal company procedures for the formal and public applications for statutory permits.

At all the stages described in Chapter 4, we emphasise the benefits of informal consultation with the local planning authority, the Environment Agency and any local interests that are likely to be affected by the project.

Figure 2A Project development - traditional

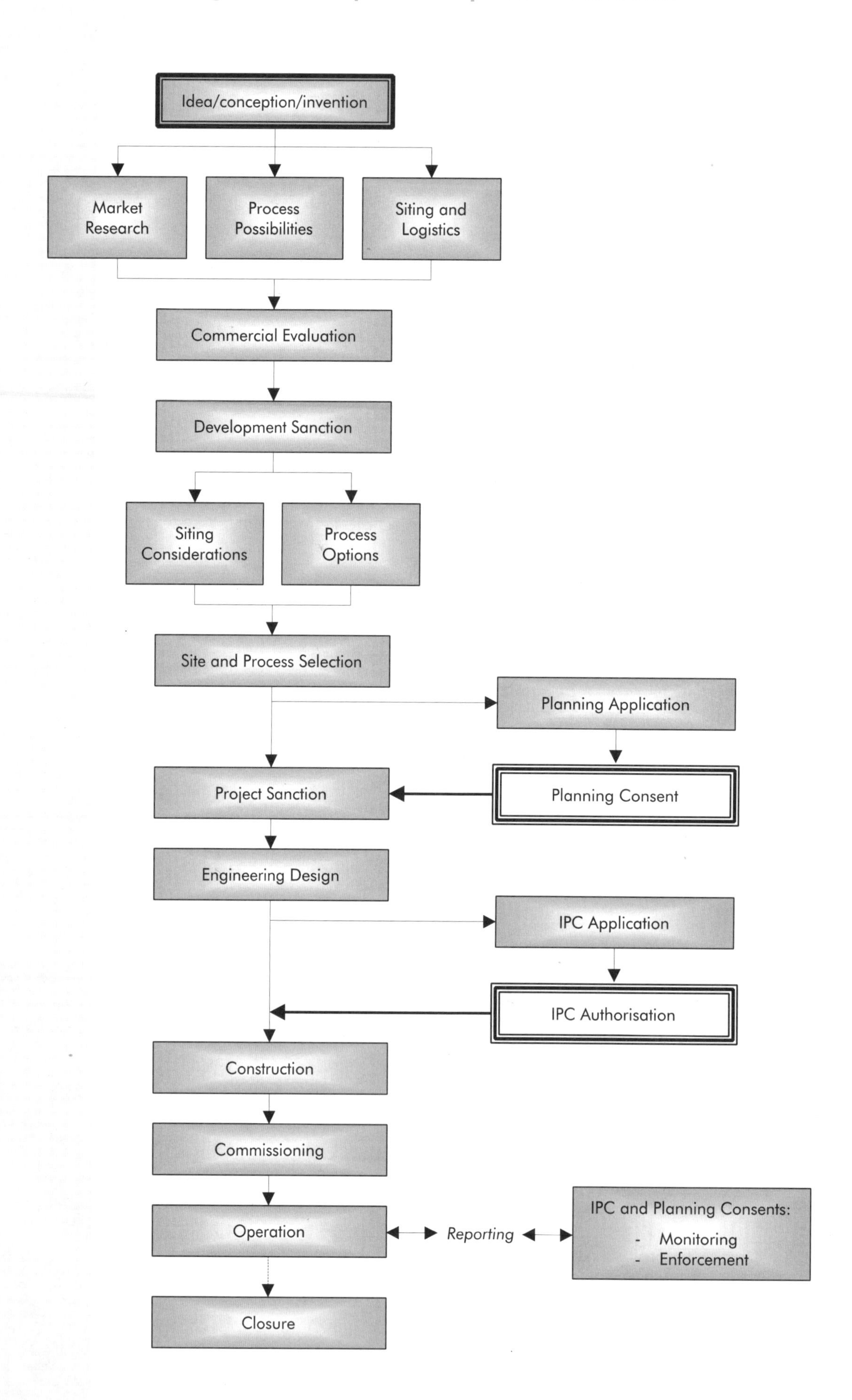

Figure 2B Project development - recommended

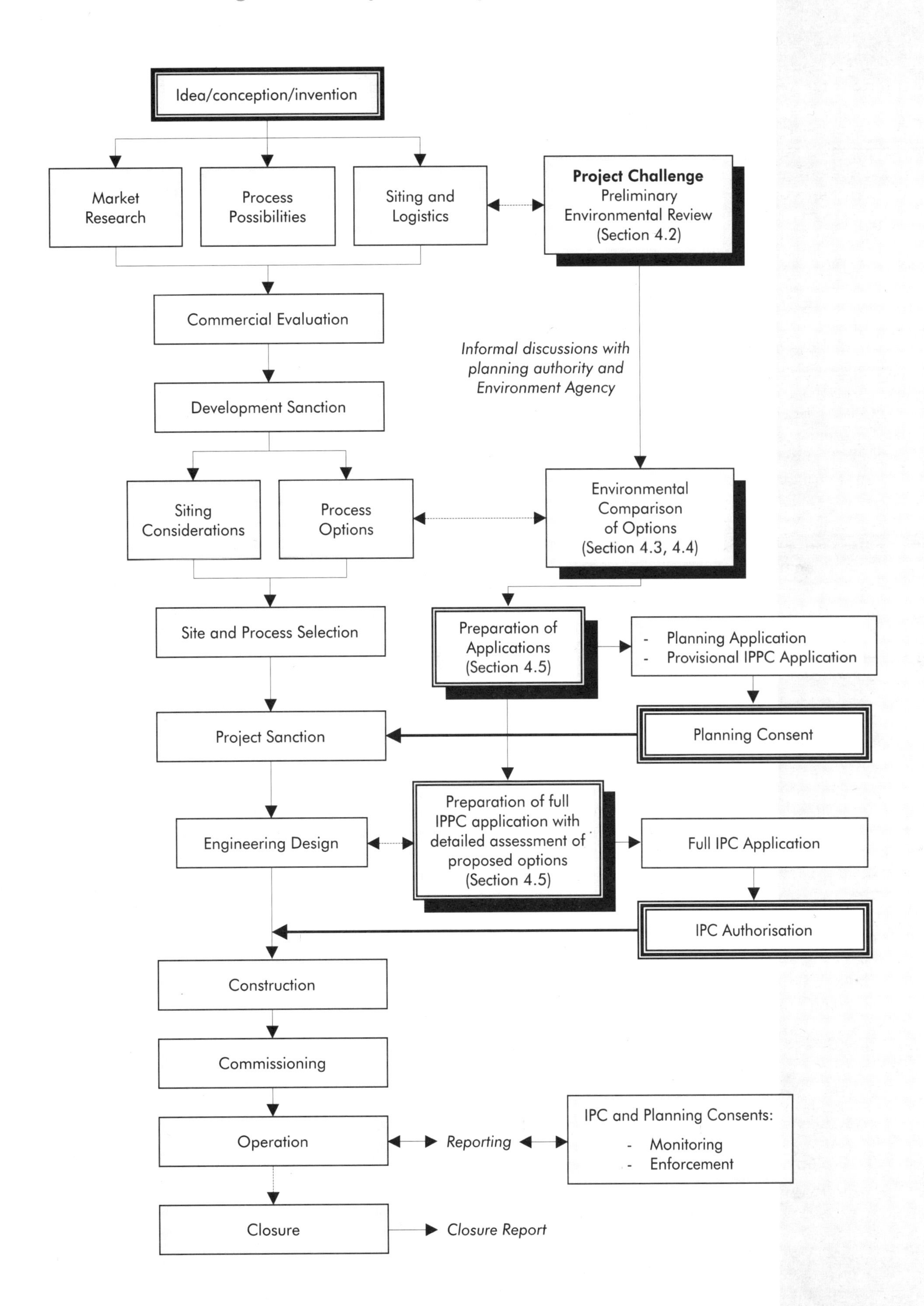

3. PLANNING AND POLLUTION CONTROL

3.1 Planning consent

Planning consent is required for most developments or material change of use of land. The purpose of town and country planning is to protect amenity and the environment in the public interest. In this context the public interest includes social and economic considerations as well as environmental aspects. A planning authority's assessment of an application for development will take into account the consequences of the proposal for matters such as employment, public rights of way and other amenity aspects as well as direct environmental impacts such as pollution and noise. The planning authority will also consider how the proposed development might affect other activities in the area and whether it is consistent with any local development policies and plans. Local development plans generally specify areas where particular types of development would not be permitted.

These social and economic considerations carry considerable weight when a local planning authority is considering a proposed development. As the developer, you should make sure that you have taken account of all these issues and understood their implications for your plans at the earliest stages of your project development programme.

One of the main considerations for the planning authority, in determining a planning application, is to ensure that the development would not cause any harm to human health or the environment. Information about emissions from the proposed process is therefore relevant for planners. The planning application must include sufficient information to enable the planning authority to decide that the development would be acceptable at the proposed location. Precise figures and details of plant to be used are not as important as an understanding of the total emission load and potential impacts. The planning authority will also want assurance that the proposal would not change materially after permission was granted. Any significant change in the development may require a new planning application to be submitted.

Certain large-scale developments are subject to Regulations (Ref. 2) implementing the EC Directive (85/337EEC) on Environmental Impact Assessment (EIA). The Directive defines the scope of environmental assessments that must be prepared for such projects. The Department of the Environment has published guidance (Ref. 3) on the "evaluation of environmental information" for projects subject to the EIA Directive. The guidance in this book is intended primarily for managers of smaller scale projects which would not be subject to the EIA Directive but the techniques we recommend are applicable for any scale of development.

The scope of Directive 85/337/EEC is extended by Directive 97/11/EC and the government is preparing new regulations to implement the new Directive. The majority of IPC/IPPC processes will still be below the thresholds requiring statutory environmental impact assessments and statements.

If an application is refused planning consent, the applicant has the right to appeal and the appeal can be determined following written representation, an informal hearing or a public inquiry heard by an inspector. The public has the right to be heard at an informal hearing or a public inquiry. Members of the public also have the right to make written representations.

The Secretary of State or the Planning Inspectorate makes the final decision on appeals, not the local planning authority.

References

2. The Town and Country Planning (Assessment of Environmental Effects) Regulations 1988, SI 1988/1199 and Environmental Effects (Scotland) regulations 1988, SI 1988/1221.

3. Evaluation of environmental information for planning projects - a good practice guide, 1994, Department of the Environment.

3.2 IPC/IPPC authorisation

IPC was introduced by the 1990 Environmental Protection Act. This requires specified categories of industrial processes to use "best available techniques not entailing excessive costs" (BATNEEC) to prevent or minimise and render harmless releases to the environment. It also requires such processes to minimise pollution to the environment, taken as a whole, by "having regard to the best practicable environmental option (BPEO) available as respects the substances which may be released". This means that the operator of the process must be able to show that it has the least overall environmental impact compared with any practicable alternative ways of carrying out the process. Applications for IPC authorisations are advertised and the public has the right to comment. IPC is administered by the Environment Agency but local authorities are statutory consultees on all IPC applications. There are arrangements for appealing against an Environment Agency decision not to authorise a process, but the IPC procedures do not provide opportunity for a public inquiry.

The 1995 Environment Act places a duty on the Environment Agency to have regard to sustainable development in carrying out its regulatory functions. The government has not issued guidance on how this duty would be implemented. However the procedures recommended in this guidance should help managers of new industrial developments to demonstrate that their proposals would have the least practicable environmental impact.

New UK legislation will be introduced in 1999 to implement the EC Directive on Integrated Pollution Prevention and Control (IPPC). The requirements of IPPC for pollution control are broadly in line with the BATNEEC and BPEO requirements of IPC, but the Directive also requires consideration of more indirect environmental effects, including energy consumption, waste minimisation, accident prevention and site restoration. This will bring the scope of environmental regulation closer to the concept of sustainable development. The IPPC Directive covers all the processes currently subject to IPC. It also applies to a wider range of manufacturing and agricultural activities.

3.3 Linkages and overlaps between Planning and IPC/IPPC

From the brief outlines given in sections 3.1 and 3.2, the primary purpose of the land use planning system can be stated as answering the question, *"is this an appropriate location for a development of this type?"* The primary purpose of IPC/IPPC can be stated as answering the question, *"given a development of this type in this location, what controls must be incorporated to achieve a high level of protection of the environment?"*

Expressed in this simplified fashion, it is clear that both the planning authority and the IPC/IPPC regulator require information about the actual and potential environmental effects of a proposed development before they can make their respective judgements and reach decisions to accept, reject or seek modification to a proposal.

Land use planning is primarily a process carried out on behalf of the local community, although within a framework of legislation and national guidance. Planning decisions must take account of local interests and concerns, including social and economic considerations as well as direct environmental impacts, such as pollution and noise. IPC/IPPC authorisation takes account of nationally based guidelines on BATNEEC (or BAT as defined in the IPPC Directive) for the particular category of manufacturing process. The conditions of an authorisation must also ensure that permitted releases and techniques for minimising accidental releases ensure adequate protection for public health and the local environment. The Environment Agency will also consider the contributions that IPC/IPPC authorised processes make to any national targets and programmes for limiting pollutant releases, e.g. the national target for reducing sulphur dioxide emissions.

In the Introduction and Figure 1, we emphasised that land use planning takes account of broader environmental effects than IPC/IPPC. Another distinction between the two systems is that planning consent relates to the whole life span of a project, from construction to end-of-life site clearance whereas IPC regulates only the operating phase. The IPPC Directive is also mainly concerned with the operating phase but it includes a new requirement for measures to avoid any pollution risk remaining after the activity has closed down.

Another important distinction between the two systems is that local political considerations can, quite properly, influence the planning process. Local structure plans usually designate areas for particular categories of use or amenity. Protection of these areas may also constrain industrial development in neighbouring areas. These aspects are not easy to assess or compare in a direct quantitative way, but as a developer you must take them into account in your project plans. You should do this in as structured a way as possible. Guidance on these aspects is included in Chapter 4.

4. ENVIRONMENTAL ASPECTS IN PROJECT DEVELOPMENT

4.1 Introduction

This is the main part of this guide. It sets out a procedure for incorporating environmental considerations into your project development programme to ensure that you take account of all the environmental aspects of your proposed process from the earliest stage of your project plans. The key parts of the procedure are the highlighted boxes in Figure 2B. Figure 3 gives more detail of the steps involved.

The procedure begins with a preliminary environmental review. This should be planned as soon as the business opportunity has been identified. This is to ensure that all significant environmental aspects that might affect the viability of the project are identified before any major commitments are made. It also provides initial information about the environmental aspects that are likely to require more detailed evaluation as the design of the project develops. The main purposes of this preliminary review are:

- To challenge whether the proposed project could satisfy all the environmental concerns that might arise;
- To reduce the risk of surprises at later stages in the project when changes could involve delay and expense.

It also serves as an initial screening stage of the possible options that you may be considering for implementing the project. For example, if a project is likely to produce an effluent that would not be acceptable for discharge to sewer, it is better to be aware of this before detailed design work is commissioned. The technical and economic consequences of needing in-house effluent treatment could jeopardise the viability of the whole venture.

The procedure for carrying out the preliminary environmental review is described in section 4.2.

The next stages of the procedure are:

- examination and assessment of the environmental aspects of process options;
- examination and assessment of the environmental aspects of site options.

These are closely inter-linked and must be considered in parallel. Section 4.3 describes techniques for comparing the environmental effects of options.

If the option with the best overall environmental performance is significantly more costly than the next best alternative you will want to be satisfied that the higher cost is justified in relation to the environmental benefit it produces. Section 4.4 gives some advice on the difficult question of cost-benefit assessment.

Having decided on the preferred process and location, the next stage is to prepare a detailed environmental assessment of the proposed development to demonstrate that it would meet all the relevant environmental regulations and be environmentally acceptable at the proposed location.

This staged procedure should ensure that you obtain and assess all the information you will need to prepare the formal planning application and IPC/IPPC application at the appropriate time in your project development programme. It should also ensure that you consider all the environmental issues that are likely to concern the authorities and others who may be affected by the project and take them into account as you develop the final design of your new process. By following this procedure you will be able to display and justify the reasons for your process and site selection decisions to the regulating authorities and thereby minimise the risk of delay in obtaining your planning permission and IPC/IPPC authorisation.

At each stage of the procedure we provide checklists and other techniques to help you identify all the potential environmental impacts of your project. We also make recommendations about consultation with authorities and others during your project development programme.

For clarity of explanation, the procedure we have outlined above and describe in more detail in the following paragraphs, draws a line between the preliminary environmental review and the subsequent assessment of project options. In practice the distinction is often not so clear cut. Nor is it our intention to suggest that there should always be a number of clear-cut options to assess. What is important is to have an approach to project development that promotes the generation and critical evaluation of options at each stage in your project development programme, from inception through to detailed engineering design. Your objective is to build up a clear and reasoned justification for your project that will stand critical review. The procedures we describe in this guide are simply tools to be used and adapted to help you achieve that objective.

Chapter 4 concludes with advice in section 4.5 for preparing environmental reports to support your formal applications for planning consent and IPC/IPPC authorisation.

Figure 3 Environmental assessment throughout project development

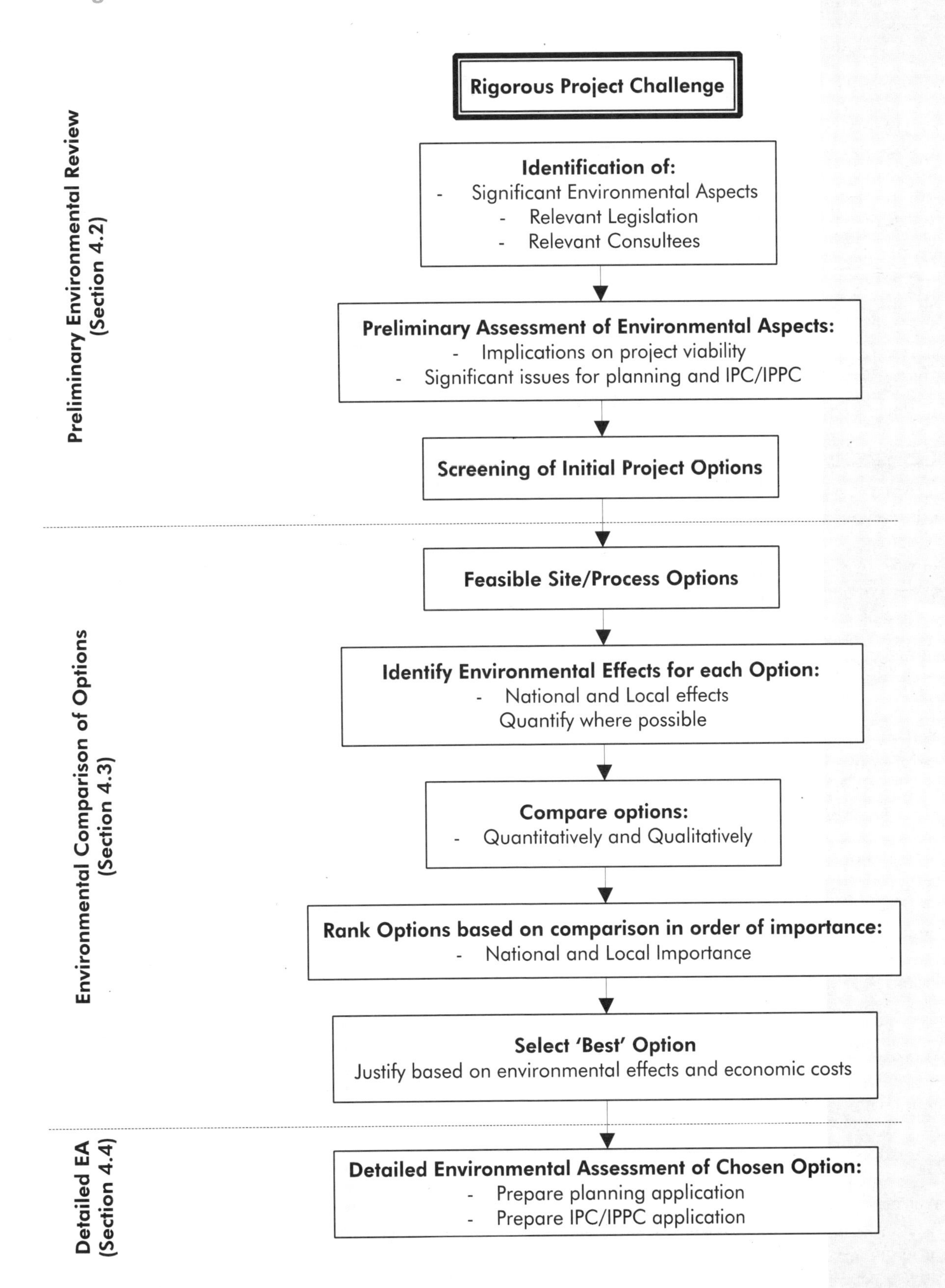

4.2 Preliminary environmental review

As soon as you, or your company, has decided that a new business opportunity is likely to be worth pursuing you should carry out a preliminary environmental review. The purposes of this review are:

- to identify the potentially significant environmental aspects of the project;
- to identify the environmental legislation that will apply;
- to make a preliminary assessment of the implications of environmental aspects on the technical and commercial feasibility of the project;
- to identify the environmental aspects that will require detailed evaluation as the design develops;
- to identify the environmental aspects that are likely to be the most significant issues for planning consent and IPC/IPPC authorisation;
- to identify authorities and other organisations who are likely to be affected by the project and who should be consulted at an early stage in project development.

The benefit of this review is assurance that you can proceed to more detailed design with confidence that you have not overlooked any significant environmental aspect.

We recommend that you carry out the review as a team exercise involving all the relevant members of the business management team and any specialist advisers. It must be very wide ranging. You must challenge every aspect of the planned project to test whether all environmental effects have been identified and fully considered. There is always a danger that enthusiasm and eagerness to commit to a particular project could be a temptation to underplay important environmental issues. This could leave your business exposed to difficulties and costly delays and changes at a later stage, when those aspects are challenged by the authorities or other interested parties.

You should carry out the review in a planned and structured way, keeping a complete record of all the information you have used and the decisions you have taken. Your records should be in enough detail to enable you, or others, to review your decisions at a later date if there are questions or changed circumstances. Most of the information you compile at this stage will be needed for the more detailed environmental assessments later.

Although it is outside the scope of this guidance document, it is usually worthwhile reviewing the safety and occupational health aspects of any new development in parallel with the environmental aspects. Much of the information is common for all three aspects.

You should carry out the review as soon as possible after the potential project has been identified and before any detailed design work starts. For most projects it will have to be carried out in stages. An initial run through may reveal areas where you need more information to assess whether a particular aspect is likely to be significant.

You should also be prepared to "review the review", as the shape of the project becomes clearer. If the scope or scale of your project is changed at a later stage in the design programme you should test those changes against the information you used and the decisions you took at the initial review, to check that nothing new has been introduced that would cause undesirable environmental effects.

Figure 4 provides a checklist that can be used as the basis for a preliminary environmental review. It is a list of environment-related issues and effects that are likely to apply to most industrial projects. However it is only a *guide*. No single generic checklist can cover all the environmental aspects of all types of projects. Your review team should develop a specific checklist for your particular project.

(Annex B provides an expanded list of issues and questions under the same headings as Figure 4. This is provided for the later stages of assessing site and process options but you may also find it helpful for compiling the checklist for your preliminary environmental review.)

We recommend you to talk with your local planning authority and with the local office of the Environment Agency when you are preparing your preliminary environmental review. The officers can provide the regulators' views on what are likely to be the sensitive environmental aspects of your project and they can advise you of any local concerns that the project might arouse. At this stage the planning authority would normally be able to advise you on the scope and level of detail they would like to see in your assessment of the environmental effects of the project. You can then adapt your review checklist to take account of these issues. Discussions of this type, with officers, can be carried out on a confidential basis. They will give you added assurance that your preliminary environmental review has covered all the environmental issues that are likely to come under scrutiny at the later stages of formal applications for consents to build and operate.

At this stage you should also consider discussing your plans with local organisations that are likely to be affected by or concerned about your development. Aspects of commercial confidentiality may constrain such discussions but it is usually beneficial to involve those who will be affected by the project as early as possible to minimise the risk of misunderstandings at later stages.

Figure 4 Basic checklist for a preliminary environmental review (Annex B gives an expanded version of this checklist)

PROCESS CONSIDERATIONS

1. Substances - type, quantity, raw materials, products, by-products, wastes, packaging
2. Energy - fuels, power supplies
3. Fresh water consumption
4. Emissions to air - type, quantities
5. Liquid effluents - type, quantities, disposal methods
6. Wastes to land-based disposal - type, quantities, disposal methods
7. Hazards - CIMAH criteria, other hazards
8. Noise and vibration
9. Lighting
10. Visual appearance - areas, heights, location, types of structures
11. Transport - materials, people

SITE CONSIDERATIONS

12. Present use - condition
13. Surroundings - sensitive sites
14. Environmental quality - pollution, ecology, amenity
15. Local Development Plans

LEGAL CONSIDERATIONS

16. Legislation - relevant authorities, local officers

ECONOMIC & SOCIAL CONSIDERATIONS

17. Employment
18. Supplies and services
19. Local concerns

When you have checked (or challenged) your project against all the relevant aspects on your checklist, you should have the following information:

- a note of any environmental issues which could materially affect the technical or commercial viability of the project;
  If the review has revealed a potential "show-stopper" you have no alternative but to reconsider the whole scheme and seek an alternative approach;
- a list of all the environmental aspects that will require detailed evaluation as the project design is developed;
- a note of the environmental aspects which are likely to be major issues for planning and/or IPC/IPPC applications;
- a note of the applicable environmental legislation and names of the local officials responsible for enforcement;
- a note of any local concerns and interests that should be taken into account, and any consultations that should be planned in advance of formal applications.

You should record these findings as part of the project documentation and use the information to guide the next stage, described in section 4.3.

Depending on the extent of any initial consultations you have had with your local planning officer and Environment Agency inspector, it may be worthwhile discussing the outcome of the preliminary environmental review with them. They will not be able to endorse the findings in any formal sense and there are no procedures for formal approval at this stage. However, they will usually be pleased to discuss and advise on any aspect they think has been overlooked or not given sufficient consideration.

The key to a successful preliminary environmental review is a rigorous challenge to the process at the beginning, questioning every aspect of the proposed development that could have environmental effects. The cost of change is least at the start of a project and becomes increasingly difficult and costly as the project becomes more developed.

4.3 Project options - assessment of site and process options

Provided that your preliminary environmental review has not revealed any potential problems that would cause you to reconsider the whole project, the next stage is to examine the practicable options that are available to you for implementing the project. Both the planning authority and the Environment Agency will require evidence that your final design of the project is the best practicable scheme on environmental grounds. The best way for you to do this is to show that you have considered all reasonable alternatives and why they would have less favourable environmental effects than your proposed project.

The information generated by the preliminary environmental review will serve as a screening stage to eliminate some options that are not worth further consideration because of insurmountable environmental constraints. For example, if one option would produce an effluent requiring on-site waste water treatment and there was no space for such a facility on the only available site, there would be no point in examining that option in any more detail.

4.3.1 Identifying practicable options

The scope for examining options varies considerably from one project to another. In some cases there may only be one technical process for a particular manufacturing operation; in others there may be several choices. Even where there is little choice in the basic manufacturing process there may be options in supporting services, e.g. energy supplies or waste treatment and disposal.

You may or may not have a choice of sites for the proposed project. Where there are realistic choices of sites, your applications for consents should include justification for the proposed site. Site selection can refer to geographically separate sites or to the precise location of a proposed facility within a large site.

The environmental effects of an industrial operation depend on the processes carried out and their location. For this reason you need to consider process and siting options together when compiling and comparing environmental effects. For example, if there are two practicable manufacturing processes, P1 and P2, and two alternative sites, S1 and S2, the options that would have to be assessed and compared are:

1. P1 at S1
2. P1 at S2
3. P2 at S1
4. P2 at S2

Likewise, if there are practicable alternatives for waste treatment and disposal within any of the process-site pairings that would lead to a further series of options for consideration.

We use the word "practicable", with reference to selection of options, to mean options that are technically and commercially viable. There is no point in examining purely theoretical or otherwise unrealistic possibilities - but the onus is on you, as the developer, to satisfy the regulating authorities that you have considered all practicable options.

At the beginning of the procedure for examining process and site options you should identify all practicable combinations. Arbitrary elimination of options at this early stage risks missing a potentially best choice that may not have been obvious at the outset. It could also leave gaps in the justifications for the planning and IPC/IPPC applications leading to costly delay at a late stage in your project development programme.

In practice the number of options that you have to examine is generally limited to a small number by commercial and other business considerations. For example, many companies will not have any choice of site for a new plant, other than their present site, because the proposed new plant would be dependent on existing site services. In many instances your choice of manufacturing process may be restricted by contract arrangements or your customers' requirements.

4.3.2 Identifying and quantifying environmental effects

Having identified your practicable options, your next step is to identify all their environmental effects, in quantitative terms as far as possible. At this stage each option should be developed to at least a block flow diagram with material flows and utility requirements (e.g. power and water supplies). Mass and energy balances are a useful check that no by-products or waste streams have been overlooked. Designs should be developed in sufficient detail to be sure that the process could meet current and anticipated regulatory requirements for emission limits and environmental quality limits. There is no point in considering options that could not meet legal requirements.

"Released Substances and their Dispersion in the Environment" (Ref. 1) gives guidance for compiling emission inventories and estimating dispersion in the environment and resulting concentrations of substances in the local environment. At this stage of your project development programme it will normally be sufficient to use very simplified methods for any dispersion calculations. We are only looking to rank the overall environmental effect of each option at this stage - a detailed and fully quantitative environmental impact assessment of each option is not required.

The recommended procedure is:

(a) Identify and quantify (as far as possible) all the environmental effects of each option. (The checklist in Annex B will be useful at this stage.)

(b) Check whether this more detailed analysis reveals any significant environmental effects that were not identified in the preliminary review. If so, revisit the preliminary environmental review findings and check whether the new information would affect the viability of the project.

(c) Display the results from (a) in a table to facilitate comparison between options.

The outcome of this procedure should be a table showing all the significant environmental effects for each option. Figure 5 illustrates this for a project with two process options and two site options. The aim of the table is to display corresponding environmental effects to permit comparison between options for each type of effect. The full list of descriptors of environmental effects in the table for your particular project will be specific to the processes and sites that you are considering, The output of your preliminary environmental review gives you a starting list of significant environmental effects and Annex B gives a checklist of environmental issues and questions which you can use as a prompt for identifying more detailed environmental effects of your project.

Note that Figure 5 shows "environmental effects" in terms of both releases from the proposed development and the resulting levels of contamination of the local environment, e.g. tonnes of sulphur dioxide emitted and estimated ground level concentration. This is necessary for the methods, described in the following paragraphs, for comparing the environmental effects of options.

The environmental effects, which are considered in land use planning and IPC/IPPC, are those resulting directly from the proposed development at the selected site. The environmental effects of producing the raw materials and services used in the process do not have to be assessed. However the IPPC Directive introduces a requirement to demonstrate that "energy is used efficiently". To cover this aspect we recommend that you determine the carbon dioxide produced in providing all the energy used by your process and include this in the comparison table as an environmental effect. This is the most convenient measure of your project's contribution to the "greenhouse effect", which is the principle environmental consequence of energy consumption. You will need to estimate the carbon dioxide produced in the generation of any electricity purchased from the national grid as well as the carbon dioxide emitted from any fuel burning within your process.

While upstream and downstream environmental effects are not material considerations for planning or IPC/IPPC, it is good environmental management practice to consider the environmental effects of the whole supply chain when planning the development of a new product or service.

You should also include some measures of the environmental effects of any on-site or off-site disposal of wastes from your proposed process, e.g. tonnes to landfill, but you do not need to assess the environmental effects of product distribution, use or disposal for IPC/IPPC or planning applications.

We recommend that you use the same table to display and compare the economic and social effects of your process and site options. The "employment and local economy" entry in Figure 5 shows an example where quantification is possible in this area.

For some economic and social effects it is not always possible to use quantitative measures. The last two entries in Figure 5 are "acceptability of new industry" and "visual impact". These are subjective aspects of the local community's attitude to the proposed development. Here we use qualitative assessments - *high/medium/low* - based on local knowledge and experience. The discussions you had with the local planning authority at the preliminary environmental review stage should be helpful in this area. In spite of the imprecise nature of the basis for comparison, issues of this type can be crucial considerations for a new development and should always be included in your overall comparison of options.

Figure 5 Comparing environmental effects of project options

Environmental effect (or load or impact)	**Option A** (P1 at S1)	**Option B** (P2 at S1)	**Option C** (P1 at S2)	**Option D** (P2 at S2)	**Comparison** better	worse
A) Emissions to air						
Sulphur dioxide - release (tonnes/yr)	190	5	190	5	B/D	A/C
- max. glc (15 min.) (ppb)	30	2	30	2	B/D	A/C
Nitrogen oxides - release (tonnes/yr)	60	80	60	80	A/C	B/D
- max. glc (1 hr) (ppb)	5	7	5	7	A/C	B/D
Air quality - current						
Sulphur dioxide - 15 min (ppb) current (plus project)	20(50)	20(22)	50(80)	50(52)	B A	D C
Nitrogen dioxide - 1 hr. (ppb) current (plus project)	100(105)	100(107)	250(255)	250(257)	A B	C D
b) Air quality objectives - for comparison with above						
Sulphur dioxide - 15 min. (ppb)	100	100	100	100	All within	
Nitrogen dioxide - 1 hr. (ppb)	150	150	150	150	A/B ok	C/D not ok
c) Liquid effluents						
COD (tonnes/yr)	1.5	2.5	1.5	2.5	A/C	B/D
Heavy metals						
- chromium (Kg/yr)	22	30	22	30	A/C	B/D
- nickel (Kg/yr)	15	20	15	20	A/C	B/D
River water quality - current plus project (River Ecosystem classification)	RE3	RE3	RE5	RE5	A/B	C/D
d) River water quality objectives - for comparison with above	RE3	RE3	RE4	RE4	A/B ok	C/D not ok
e) Energy consumption						
Carbon dioxide equivalent (tonnes/yr)	230,000	190,000	230,000	190,000	B/D	A/C
f) Transport						
Heavy vehicles (movement/day)	35	40	20	25	C D	A B
Light vehicles (movements/day)	100	100	50	50	C/D	A/B
Houses affected (number)	25	25	100	100	A/B	C/D
g) Employment and local economy						
New jobs (number)	25	20	15	10	A B	C/D
Local services used (£/yr)	£200,000	£150,000	£300,000	£300,000	C D	A/B
h) Acceptability of industrial development	Low	Low	Medium	Medium	C/D	A/B
i) Visual impact	High	Medium	Medium	Low	D B/C	A

This is a hypothetical example to illustrate the procedure.
It assumes two process options, P1 and P2 and two site options - S1 (a rural site) and S2 (an urban site.)

4.3.3 Comparing the overall environmental effects of options

The environmental effects of a development can be considered in two categories:

- national effects, expressed as contributions to national environmental objectives, e.g. annual releases of acid gases and greenhouse gases to air or persistent substances to water, for which national reduction targets have been set;
- local effects, e.g. effects on local air and river water quality or noise or traffic. Economic and social effects are also local for all but very large projects.

In regard to emissions from the development, local effects are relevant considerations for both land use planning and IPC/IPPC. Contributions to national effects will be more significant for IPC/IPPC. This is why the environmental effects in the example shown in Figure 5 include total annual discharges into the environment (environmental loads contributing to the national inventory) as well as estimated effects on local environmental quality.

For each environmental effect the relative environmental merit of each option can now be assessed. The comparisons made in the right hand column of Figure 5 show the respective ranking of each option against each environmental effect.

If this initial analysis shows that one option is the best in respect of all the significant environmental effects there is no need for further assessment. That option is clearly the best environmental option and you can proceed to detailed engineering design.

In many cases the comparisons column will not show such a clear conclusion. The example in Figure 5 illustrates such a situation. The following paragraphs describe a procedure for ranking options when their relative environmental effects vary in this way.

4.3.4 Ranking options on the basis of environmental effects with national implications

There are several categories of substances that are recognised internationally as contributing to long term harm to the global environment.

- "greenhouse gases", such as carbon dioxide (from fossil fuel burning) and methane (from mining and agriculture), which are increasing in concentration in the atmosphere and believed to be causing global warming;
- ozone depleting substances, such as CFCs, which destroy the stratospheric ozone layer, leading to increased levels of harmful ultra-violet radiation at the ground;
- acid forming gases, such as sulphur and nitrogen oxides, which lead to acidic rainfall, damaging plants and aquatic life in lakes and rivers;
- volatile organic compounds (VOCs), such as solvent vapours, which contribute to elevated concentrations of ozone at ground level, often far distant from the source, with damaging effects for human health and vegetation;
- heavy metals and persistent, toxic and bio-accumulative organic compounds in discharges to water courses, leading to long term contamination of the marine environment and harm to marine life.

Various international bodies have developed, or are working on, international agreements aimed at reducing global releases of these substances. These agreements will lead to national emission reduction targets. While there is no scientific basis for assuming that all the targets have equal environmental importance, they will be part of the UK's environmental improvement programme and that fact can be taken as an arbitrary indicator of their environmental importance. On this basis, the contribution that the emissions from a process make to the UK targets can be used to assess the relative environmental importance of those emissions.

National emission targets have not yet been set, or proposed, for all categories of substances that contribute to long term environmental damage. For example, national targets have not been set for discharges of metals and persistent, toxic and bio-accumulative (ptb) organic compounds to the marine environment. As an alternative measure of national significance, the present national emission totals may be used to compare the relative environmental effects of options.

	Current (1995) UK emissions (tonnes)	UK target (2010)* (tonnes)
CO_2	148 million	127 million[1]
SO_2	2,365,000	279,000[2]
NO_2	2,295,000	753,000[2]

(1) 80% of 1990 level (2) EC proposed strategy.

** These figures are not official UK policy but are provided as a reasonable basis for the comparison procedure recommended in this guide.*

UK dischargee of harmful substances to marine waters in 1995 (tonnes)

Cadmium	21
Mercury	4
Copper	630
Lead	390
Zinc	2800
Arsenic	85
Chromium	250
Nickel	300
PTB organics	150
Pesticides	10

Source: DETR Digest of Environmental Statistics, No. 19 1997 (means of lower and upper estimates)

In Figure 5 the following effects can be identified as "national effects":

- Tonnes per year of sulphur dioxide, nitrogen oxides and carbon dioxide released to atmosphere.
- Kgs per year of heavy metals released to water.

Figure 6 shows how the releases of these substances can be expressed as the process' contribution to national annual targets or totals.

Figure 6 Expressing releases as contributions to national annual targets or totals

Substance	National Target or Total (NT) (tonnes)	Option A emissions		Option B emissions		Option C emissions		Option D emissions	
		tonnes	% of NT	tonnes	% of NT	tonnes	% of NT	tonnes	% of NT
To air:									
Sulphur Dioxide	279,000	190	0.07	5	0.022	190	0.07	5	0.002
Nitrogen Dioxide	753,000	60	< 0.01	80	0.011	60	0.007	80	0.011
Carbon Dioxide	127 million	230,000	0.18	190,000	0.15	230,000	0.18	190,000	0.15
To water:									
Chromium	250	0.022	< 0.01	0.030	0.01	0.022	< 0.01	0.030	0.01
Nickel	300	0.015	< 0.01	0.020	< 0.01	0.015	< 0.01	0.020	< 0.01

The figures under " % of NT" can be used as an indicator of the relative significance of each release:
(a) in relation to each other;
(b) in relation to the UK's environmental improvement programme.

In this example, if 0.1 % of NT is taken as a "significant" release in the national context, carbon dioxide is the only release of national significance and options B and D are preferable on this basis. However the difference between the carbon dioxide releases of B and D compared with A and C is relatively small, so this aspect should not carry much weight in the overall comparison of the options.

Substances are in these "NT" categories because, cumulatively, they contribute to long term damage to the global environment. They are, therefore, important indicators for sustainable development. This may be a key consideration in the final selection of the project option in a situation where consideration of national effects points in favour of one option and consideration of local effects points to another.

4.3.5 Ranking options on the basis of local environmental effects

The environmental effects of pollutants in the NT categories, described in the previous section, are due to accumulation of the substances, or their degradation products, in the environment over long timescales, leading to long term and possibly irreversible effects on the natural environment. At the local level, i.e. close to the point of release, the environmental effects of releases that concern us are the possible consequences of direct exposure of people or plant or animal life to concentrations of the released substances in air, water or ground. The aim of environmental quality management at the local level is to ensure that concentrations of substances in air, water or the ground are kept below levels that would cause harm to people or flora or fauna.

At the present time there are no local emission targets comparable to national emission targets. *(This may change in the future with the development of Local Environment Agency Plans (LEAPS), local air quality management plans and water quality objectives.)* There are quality targets for some common pollutants in air (air quality limits or standards) and for many substances in surface waters (water quality limits or standards). For these substances, comparison can be made between the level of contamination resulting from the project and the environmental quality limit. This is illustrated by the examples of sulphur dioxide and nitrogen dioxide in Figure 5. Option B is better in these respects because the resulting ground level concentrations from emissions of these two substances, combined with the existing air quality, give the biggest margin below the air quality limits, compared to options A, C or D.

Other types of environmental effects cannot be expressed in such quantitative ways but qualitative criteria can be used to make comparisons. This is illustrated in Figure 5 for "acceptability of industrial development" and "visual impact".

If one option is the best in respect of all local environmental effects that is clearly the best environmental option. If that is not the case the next step is to prioritise or rank the local environmental effects to see if that would show which option is best in relation to the higher ranked environmental effects.

Ranking environmental effects

All types of local environmental effects should be included in this ranking process, including those which can only be assessed qualitatively.

There is no unequivocal way of making such a ranking. A development will have different implications for each local interest affected by it. For example, householders living alongside the approach road to a proposed factory may rank vehicle movements as the most important effect. The local angling club will be more concerned about discharges to the river.

The process of compiling a table in the manner of Figure 5 for your project brings objectivity to the problem by quantifying effects as far as possible. It helps to put the effects of the project into perspective with established activities and environmental impacts in the locality.

Your discussions with the local authority and Environment Agency will generally provide useful background knowledge to help identify environmental effects that are of particular local concern. They should also help you to identify any local objectives and plans for environmental improvement.

When you have complied all the available information we recommend that you use the technique of force ranked pairs to put all the local environmental effects into a ranked order of importance. This is illustrated in Figure 7.

Figure 7 Technique for ranking environmental effects in order of importance

Each of the environmental effects of the project options in Figure 5 is compared in turn with every other environmental effect and ranked higher or lower. Score 1 if higher, 0 if lower. In the example below, a is compared with b, c, d, e, and f; b is then compared with c, d, e, and f, and so on. The score for each effect then indicates its relative importance in relation to the other effects and a ranked order is obtained.

Item	a	b	c	d	e	f	Score	Ranking
a		1	0	0	1	1	3	3
b	0		0	0	1	0	1	5
c	1	1		0	1	1	4	2
d	1	1	1		1	1	5	1
e	0	0	0	0		0	0	6
f	0	1	0	0	1		2	4

Each member of the project team should carry out this ranking process individually. Team members then compare results and discuss any differences in their scores with the aim of reaching a team consensus on a final ranking order. This procedure ensures that all prespectives, perceptions and opinions are made transparent and fully considered.

Another technique for ranking disparate effects is "concept ranking" whereby the environmental effects are given "importance weightings" and "significance factors". This method is described in Annex C.

You can now re-examine the local environmental effects in your table in the light of the ranking order you have compiled. The option with the majority of more favourable higher ranking environmental effects is then likely to be the option you could justify as the preferred environmental option.

This procedure for comparing and ranking options is most easily applied when the options being compared are process options at the same site. This will often be the case for smaller scale projects. Where the options involve different sites, as in Figure 5, you may find it helpful to apply the forced ranked pairs technique separately for the process options at each site if there are significant differences between the environmental priorities of each site. For example, if air pollution, noise and traffic are the main concerns at one site, but risk of water pollution is the top priority at the other.

4.3.6 Overall comparison of environmental effects

If the separate assessments of national and local effects point to conflicting conclusions, i.e. one option is preferable on national effects while another is preferable on local effects, the separate factors in each assessment have to be re-examined to reach a judgement on whether national or local effects should be determining. For most smaller scale projects, say up to £10 million capital investment, their contribution to national effects is likely to be a very small proportion of the national targets and totals and often insignificant in national terms. Local issues are then the determining factors.

Figure 8 is a diagrammatic summary of the overall procedure for selecting the preferred option.

The procedures described in the previous paragraphs and Figure 8 provide you with a structured display of all the environmental effects of each of your project options. From this information you should be able to show which option has the least effects on the environment, from both national and local perspectives. In some cases the procedures we have recommended will show you that one option is clearly preferable to any others. In other situations the procedures may not produce such a clear outcome and your final selection may still be debatable. Your final decision in those circumstances will be a matter of judgement. However the structured approach that you have followed will mean that you have identified, assessed, prioritised and documented all the key factors so that the reasons for your final choice of process and/or site can be recorded and explained to the relevant authorities and other interested parties.

Figure 8 Procedure for selecting preferred environmental option

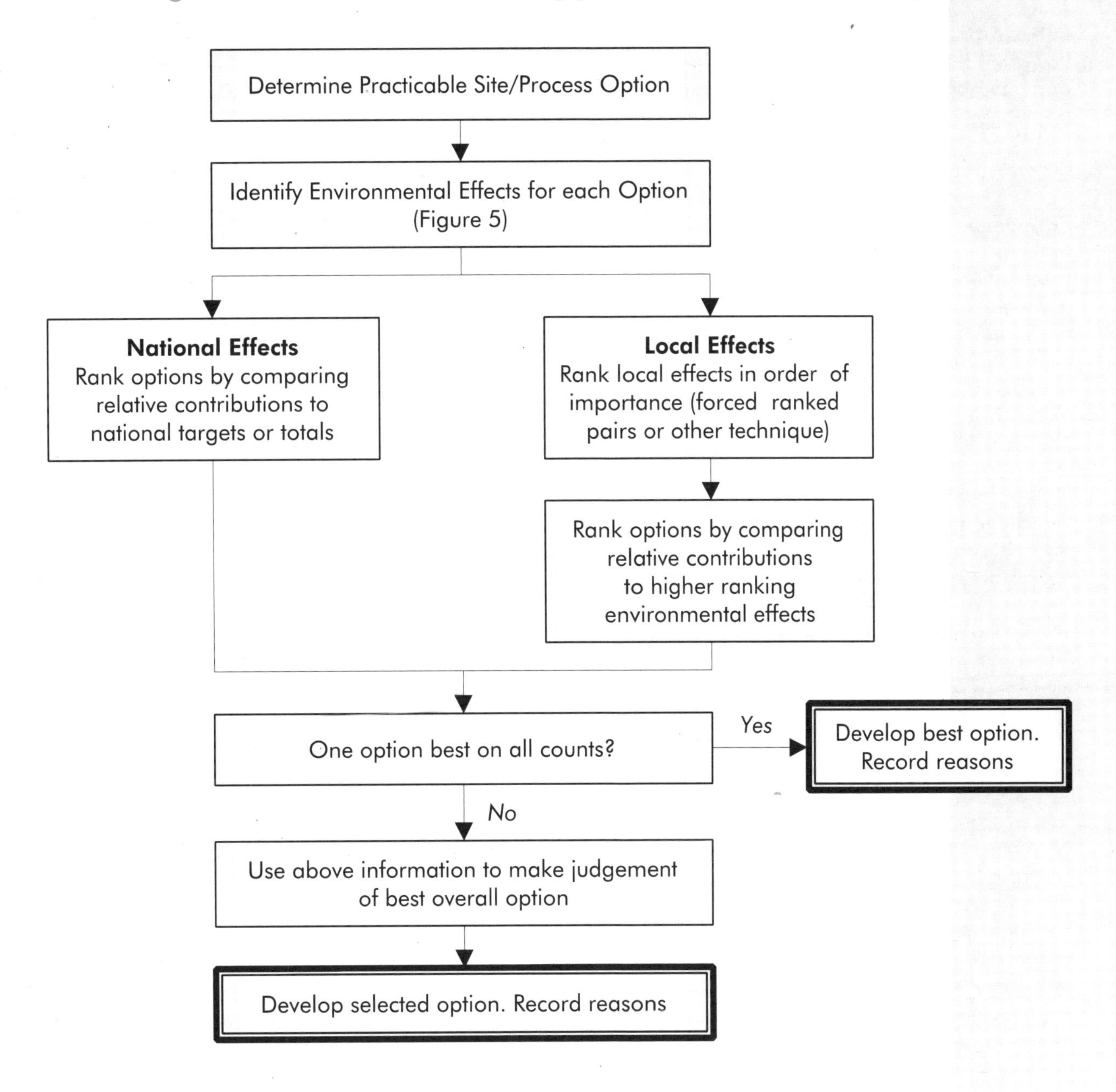

4.3.7 Recording the option selection procedure

All stages in the procedure described above should be fully documented. An account of the reasons for your choice of process and site will be required as part of the information in support of your planning application and/or IPC/IPPC application.

At this stage it is always worth considering a further informal discussion with the local planning authority and the Environment Agency to make sure that nothing has been overlooked in your decision making process that would cause questions and delays at the formal application stage.

Having identified the preferred option, you can now proceed with confidence to detailed engineering design and formal applications for planning consent and IPC/IPPC authorisation. Both applications must include an appropriate form of environmental assessment of the proposed project. This must clearly demonstrate what the environmental effects will be and show that they would be acceptable in relation to all relevant national and local requirements and standards. All the information required for both types of applications should have been compiled in the option selection procedures. Section 4.5 describes the scope of environmental assessment for each type of application and gives advice on compiling and presenting the information in an appropriate way for both types of applications.

4.3.8 Worked example

In this section we illustrate use of the procedures for comparing process options with an example based on information supplied by Pilkington plc for float glass manufacture. The data used does not apply to any particular installation and the values quoted are solely for the purpose of illustrating the procedures.

Project description

700 tonne per day of glass is produced by the float glass method. The raw materials for glass making - sand, soda-ash, dolomite, limestone and sodium sulphate - are melted in a gas fired furnace. The molten glass is then formed into a uniform layer by floating on the surface of a bath of molten tin, from where it is drawn and cooled to produce a continuous sheet of plate glass that needs no grinding or polishing. Very high temperatures are required in the melting process and the exhaust gases from the furnace contain high concentrations of oxides of nitrogen formed by oxidation of nitrogen in the combustion air. The present furnace operates within the limit of 2700 mgm / m^3 of oxides of nitrogen (NOx) in the flue gases, measured as NO_2 (corrected to 8% O_2) proposed under EPA legislation.

NOx is a major atmospheric pollutant due to its formation in all combustion processes, including vehicle engines and stationary boilers and furnaces. In combination with emissions of volatile organic compounds it also contributes to photo-chemical smog and ozone formation. More stringent regulatory limits are anticipated for all sources of NOx.

The project is to assess available techniques for modifying the existing furnace to reduce emissions of NOx. The design target is a concentration in flue gas of less than 500 mgm / m^3 measured as NO_2.

Process options

Three technically feasible options are available for gas fired glass melting furnaces that would give considerably lower NOx emissions than the present furnace.

(a) **Oxy-fuel** Pure oxygen is used in place of combustion air. By eliminating atmospheric nitrogen from the combustion chamber emissions of NOx are reduced to very low levels.

This process would require 130,000 te / yr of pure oxygen. This would be supplied, either by an on-site cryogenic air-separation plant or by pipeline from an outside supplier. In either case production of the oxygen would involve additional energy consumption to produce the oxygen by low temperature fractionation of air and to transport it to the glass melting furnace. This energy use and its associated emissions of carbon dioxide, sulphur dioxide and nitrogen oxides must be taken into account in assessing this option.

(b) **Selective Catalytic Reduction** (SCR) Ammonia is injected into the flue gas to react with NOx in the presence of a catalyst to form nitrogen and water vapour.

This process would require 600 te / yr of ammonia which would be delivered to the site and stored as anhydrous liquid under pressure. The production of the ammonia, and its transport, would involve additional natural gas and other energy supplies. This energy use and its associated emissions must be taken into account in assessing this option.

It would also be necessary to remove sulphur compounds and particulates from the flue gases to prevent catalyst poisoning. This would involve a reaction chamber and an electrical precipitator (EP) with facilities to recycle the collected EP dust.

(c) **The 3R™ Process (Reaction and Reduction in Regenerators)** Natural gas (or another hydrocarbon fuel) is injected into the hot flue gases in the regenerator system where it reacts with NOx to form nitrogen and water vapour.

Combustion of the residual fuel within the regenerator system raises the temperature of the waste gases in the flues, but this energy can be largely recovered by installing a waste heat boiler.

In all cases the flue gases would be discharged from the existing 105m stack. None of the options have any significant effect on the emissions of sulphur dioxide from the furnace since these arise from the glass making raw materials. None of the options generate any different solid or liquid wastes to the present furnace.

Comparison of options

Following the procedure from section 4.3.5 and Figure 5, the relevant data for each option are set out in Figure 9.

In this example we are comparing the 3 options against each other and against the base case of the existing furnace at the same location. In the table in Figure 9 it is therefore only necessary to consider environmental effects that differ among the options and between the options and the present furnace.

The right hand column of Figure 9 shows that Option C is better in respect of 9 of the 11 environmental effects considered. It is also the lowest cost option by a considerable margin. However, before concluding that Option C is the best overall option we should look at those effects where other options show an environmental advantage.

Nitrogen oxides - direct release from furnace and local maximum ground level concentration (glc)

Option A is better on these measures. To judge whether this is significant in relation to other measures we compare the maximum glc value for options B and C with the proposed longer term EC limit values for NO_2 and for NO + NO_2, which are 40 and 30 mg/m^3 respectively. The glc values for options B and C, of 0.12 mg/m^3, are less than 0.5% of these target limits and therefore of little significance for the local air quality.

The better performance of Option A in respect of this measure would not therefore outweigh its disadvantages in respect of the other measures where it gives poorer environmental performance.

Carbon dioxide - total release (directly from furnace and indirectly from power generation)

Option B is better than C or A on this measure. To judge whether this is significant in relation to other measures we compare the total release with the UK emissions of CO_2. As a proportion of the UK target of 127 million tonnes by 2010, the emissions from each option amount to:

Option A 0.12%;
Option B 0.11%;
Option C 0.11% (to 2 significant figures)

From these values it is clear that the advantage that Option B has over Option C is not significant.

We can therefore conclude that Option C provides the best overall environmental performance. In this example it is also the lowest cost option.

This relatively simple example shows that it is not always necessary to use the ranking procedure described in Figure 7 in section 4.3.5 because we have been able to demonstrate that the environmental measures which were less favourable for Option C are not significant in relation to recognised environmental quality criteria.

The decision would not have been so straight forward if only Options A and B were available. In that case, overall emissions to air (and costs) would favour B but hazardous materials and visual impact would favour A. In that situation the ranking procedure might be helpful.

In both cases the example illustrates the clarity that is provided by comparing all the relevant environmental effects of the options in a systematic way. A presentation in the style of Figure 9 summarises your decision making process in a transparent fashion and can be used to support your applications for IPC/IPPC authorisations or planning consents.

Figure 9 Comparing environmental effects of glass furnace options

Environmental effect	Present Furnace	Oxy-fuel Option A	SCR Option B	3-R Option C	Comparison better worse
Emissions to air					
Sulphur dioxide (direct)					
Release rate te/yr	500	500	500	500	
Max. glc $\mu g/m^3$	0.15	0.15	0.15	0.15	equal
Sulphur dioxide (indirect from additional power generation) te/yr	none	750	35	none	
Total Sulphur dioxide te/yr	500	1250	535	500	C B A
Nitrogen oxides (direct)					
Release rate te/yr	2000	< 400	400	400	
Max. glc $\mu g/m^3$	0.6	< 0.12	0.12	0.12	A B/C
Nitrogen oxides (indirect from additional power generation) te/yr	none	500	20	none	
Total Nitrogen oxides te/yr	2000	< 900	420	400	C B A
Carbon monoxide (direct)					
Release rate te/tr	175	175	175	135	
Max. glc $\mu g/m^3$	0.05	0.05	0.05	0.04	C A/B
Carbon monoxide (indirect from additional power generation) te/yr	none	250	12	none	
Total Carbon monoxide te/yr	175	425	187	135	C B A
Ammonia					
Release rate te/yr	none	none	2.6	none	
Max. glc (15 min) ppb	none	none	0.0008	none	C/A B
Carbon dioxide - direct te/yr	128,000	128,000	128,000	137,000	
- from oxygen production te/yr	none	30,000	none	none	
- from ammonia production te/yr	none	none	6800	none	
Total Carbon dioxide te/yr	128,000	158,000	134,800	137,000	B C A
Transport - change from present					
Heavy vehicle movements per yr.			+ 200		A/C B
Visual impact - change from present		New furnace Outline	EP structure	none	C A B
Hazardous materials		none	200 te ammonia	none	C A B
Net present cost £ millions		12	4	2	C B A

4.4 Costs and benefits

In section 4.3 we made the point that all the process and siting options you consider must be "practicable", meaning that they would be technically and commercially viable for your business. There is no point in examining purely theoretical or otherwise unrealistic possibilities that you would never develop under any foreseeable circumstances. We also stressed that the onus is on you, as the developer, to satisfy the regulating authorities that you have identified and considered all practicable options.

All of the options you have considered must also meet the requirements of the relevant environmental legislation. There is no point in considering options that would not be authorised. This means they must be acceptable as BATNEEC, in the case of IPC processes, or as BAT for IPPC processes. When an IPC process involves releases to more than one environmental medium the proposed operation must represent the Best Practicable Environmental Option (BPEO) with respect of the overall environmental impact of all the releases. The IPPC Directive includes a similar requirement in its reference to "an integrated approach to pollution control....wherever practicable....to achieve a high level of protection for the environment as a whole".

The 'not entailing excessive cost' element of BATNEEC refers to a balance of the environmental benefit achieved by use of the "technique" against the costs of installing and operating that "technique". However, there is no accepted method of valuing environmental benefit that would allow this balance to be determined by making direct financial comparisons. It is a question of making informed judgement. To a large extent this judgement is made by the Environment Agency (EA), after consultation with industry and others. The Agency publishes a series of IPC Guidance Notes that describe the "available techniques" and give "benchmark release levels" that new processes in each industry sector should normally achieve. The levels of environmental performance in these Notes are based on current best practice and they therefore reflect what is "practicable", in terms of technical and commercial viability, for that industry sector. The Agency expects all new processes to meet the performance levels given in the IPC Notes. The financial circumstances of an individual applicant would not justify authorisation of new processes with poorer environmental performance. However the IPC system is site specific and departures from the indicative standards may be authorised if there are technical reasons that would make some aspect of the Guidance Note impracticable in the local circumstances.

BAT for IPPC processes is likely to follow the principles of BATNEEC in incorporating cost considerations in BAT Reference Notes and in setting indicative standards.

The concept of BPEO implies a more site-specific assessment of environmental benefits and operating costs. The benefits and cost of each option should be compared and the EA will expect a clear demonstration that valid costing methods have been used in the comparisons. However, the problem of valuing differences in environmental performance between options remains.

In addition to satisfying IPC or IPPC conditions, releases from the proposed plant must not lead to any breach of environmental quality standards or compromise any national targets.

Planning controls also take into account the costs and benefits of environmental protection measures. One of the purposes of planning is to ensure viable economic development of the area concerned. The economic effects of a development, such as jobs created, will often be a central issue in determining a planning application. Therefore, in negotiations with the planning authority the costs and benefits of environmental measures and their implications for the economic effects of the development, are relevant matters.

The procedure in Section 4.3, together with the above considerations, leads to a preferred option on overall environmental effects, that is technically and commercially viable and satisfies all the relevant environmental legislation. However that is not necessarily your final position. If the option with best environmental performance is much more expensive than the next best environmental option, but the difference in environmental performance is small, then it is reasonable to question whether the higher cost is justified in relation to the environmental improvement that it gives.

This question raises, again, the problem of how to value environmental improvement (or a reduced level of risk to the environment). We are talking about valuing the difference in environmental performance between two (or more) process options which comply with all relevant environmental quality criteria. On this basis you might argue, with some justification, that there is no value in better environmental performance since, by definition, meeting the environmental quality criteria means that no harm is being caused.

It is not always the case that better environmental performance involves higher project cost.

New and innovative techniques for alternative product design or formulation, recycling, waste minimisation or energy recovery may lead to the win-win situation of lower costs with environmental improvement.

On the other hand, there is the argument that reducing levels of pollutants in the environment below the currently accepted environmental quality criteria will reduce an already low risk of environmental harm to an even lower level of risk. The justification for this reasoning is that environmental quality criteria are based on a judgement of an acceptable level of risk to sensitive targets in the environment and that concentrations below the EQC value will lead to lower level of risk. The strength of this argument clearly varies with different types of substances. It carries most weight for substances that are persistent, toxic and bio-accumulative in the environment, e.g. heavy metals and some categories of very stable organic compounds, such as polychlorinated biphenyls (PCBs). At the other extreme there are substances which are harmful at high concentrations and for which EQCs are set to limit contamination, yet they are essential at low concentrations for maintaining a healthy environment, e.g. nitrates in water.

Another reason for considering an option that goes beyond current regulatory limits is the reputation of your business. Environmental issues are now key factors in business strategy for many companies. The level of investment for a pro-active environmental policy is obviously a matter for individual business judgement.

This question raises the problem of how to value environmental improvement (or a reduced level of risk to the environment). There are theoretical economic approaches which attempt to estimate monetary values for environmental quality, but we do not think they are suitable for most IPC / IPPC applications. Our advice is to define as clearly as possible the environmental gain in quantitative and qualitative terms, set it against the costs of environmental protection measures and apply expert judgement to determine when the incremental cost exceeds the incremental gain. Regulators will not expect a full cost benefit assessment with monetary values attached to all costs and benefits but they will look for a transparent and logical application of judgement.

Another complication in any attempt to balance project costs against a value of environmental improvement is that the project costs are clearly borne by the individual or company that is financing the project, but the value of environmental improvement does not accrue to any readily identified person or organisation.

In the context of option selection for an IPC/IPPC project, it is clear that we cannot ignore cost-benefit, in spite of the lack of any direct calculus for solving the problem. What you can do is to examine the incremental costs of options in relation to measures of their environmental effects in a number of ways to provide information that will help you decide when the costs become disproportionate in relation to the environmental improvement.

(a) Plot project costs against environmental improvement.
If this plot shows an obvious discontinuity that may indicate the option with disproportionately high costs. For example, if there is a 20% increase in project cost from option I to option II for a 10% decrease in nitrogen dioxide emissions, and a 50% increase in costs from option II to option III for only a further 5% decrease in nitrogen dioxide emissions, that suggest that the extra cost of option III may not be justified by the environmental improvement it gives.

(b) Compare incremental project costs with pollution abatement costs elsewhere for controlling the same environmental effects.
If the incremental project cost is much greater than costs in other situations for reducing similar quantities of the emissions of the same pollutant then the incremental cost may be considered disproportionate. The new series of IPC Guidance Notes (and the research reports that were used in their preparation) provide information on typical pollution abatement costs for each industry sector. (Ref. 4)

We recommend that you use total annualised cost when making these comparisons.

You may find it helpful to add "option costs" to the table we used to display the environmental effects of options (Figure 5). This will complete the picture of all the information you have used to reach your final decision.

(c) More environmentally-effective alternative use of the incremental cost.
Another factor that may be relevant at this stage of your option selection procedure is whether the capital for the incremental cost of the best environmental option could be employed to greater environmental benefit elsewhere in your business. This consideration is only likely to be relevant where neither of the above comparisons leads to a clear conclusion and where the environmental improvement produced by the "best" option is small in comparison to the "next best". To justify your selection of the "next best" on these grounds you would have to link another environmental improvement project with the subject project and satisfy the regulating authorities that the combination was best for the local environment.

Whenever you make the decision that the cost of the project option with the most favourable environmental performance is disproportionate to the extra cost over the "next best" option, you will have to justify your decision to the regulating authorities.

References

4. IPC Guidance Notes (prepared by the Environment Agency) and Chief Inspector's Guidance Notes (prepared by the former HMIP), available from Stationary Office bookshops and accredited agents.

4.5 Compiling and presenting environmental assessments

The differences and overlaps between land use planning and IPC/IPPC were shown in Figure 1 in the Introduction and explained further in Chapter 3. From this it follows that there will be differences in the scope and presentation of the environmental assessments that you need to prepare to support each type of application.

Another difference between planning and IPC/IPPC is the question of timing. A developer will usually seek planning consent, and thereby assurance that the project can proceed at the selected site, before committing expenditure to the detailed engineering design that must be well advanced to provide the technical detail required for IPC/IPPC application. The diagrams we used in Figures 2A and 2B, to illustrate typical project development programmes, illustrate this point. This timing question has implications for the amount of detail that is available and necessary for both types of application.

In Figure 2B we show a "provisional IPPC application" that would coincide with the planning application. The intention of this procedure is that the Environment Agency would be in a position to give approval in principle on the basis of the information that was available at that stage of the project development programme. This would provide added clarification and assurance to the planning authority that the final IPPC authorisation, at a later date, would provide all the necessary environmental protection safeguards to prevent harm to people or the local environment.

In its response to consultation papers on implementation of the IPPC Directive, the government has indicated that it is likely to include arrangements for the Environment Agency to issue provisional IPPC authorisations in new legislation to implement the Directive in 1999.

Planning applications for large scale or sensitive developments, as specified in the Regulations for implementing the EIA Directive, must be accompanied by an "environmental statement". The Department of the Environment (now DETR) has published guidance for carrying out the environmental assessments and compiling environmental statements for such developments (Ref. 5). Applications for IPC/IPPC must show that the project will meet all relevant pollution control requirements, but there is no formal requirement for a separate "environmental statement".

The term "environmental statement" is generally taken to refer to the formal environmental assessment statement that is a legal requirement under the UK regulations implementing the EIA Directive. To distinguish between that type of formal statement and the environmental information that this guide recommends to support any planning and IPC/IPPC applications, the term "environmental report" is used in the following paragraphs.

In Part 3 we have examined the distinction between planning, as the process for deciding, *"is this an appropriate location for a development of this type"*, and IPC/IPPC as the process for deciding, *"given a development of this type in this location, what controls must be incorporated to achieve a high level of protection of the environment"*. You should keep this distinction in mind when you are preparing the respective environmental reports in support of each type of application. As a general rule, in regard to emissions and discharges to the environment, the need at the planning stage is for an understanding of the upper bounds of any likely pollution levels. Technical details of control measures and detailed estimates of environmental concentrations are matters for IPC/IPPC regulation. However, for larger developments and those in sensitive locations more detail is likely to be required at the planning stage. For those environmental effects of a development that are not regulated under IPC/IPPC, such as effects of traffic and visual impacts, you will be expected to present full information to support your planning application.

We recommend that you prepare an environmental report for all industrial projects, whether or not the project is subject to the EIA Regulations. A comprehensive environmental report provides the information that will demonstrate to the planning and pollution control authorities that you have considered all the environmental aspects of the project. It will reduce the risk of delays in processing your application. If you have followed the procedures we recommended in Part 4, you will already have all the information you need for preparing the environmental report(s). The exercise of preparing an environmental report is your final check that:

- all significant environmental issues have been considered;
- all significant environmental impacts have been assessed;
- the project will meet all environmental regulations;
- all anticipated concerns have been addressed.

It is good environmental management practice.

4.5.1 Environmental reports for planning applications

Scoping

The informal discussions you had with local planning officers, at the stages of preliminary environmental review and process and site selection, should have helped you identify all the environmental effects which are likely to be significant for your project. However, before you start to prepare the environmental report to accompany your planning application, it is usually worthwhile checking with them again on the scope (i.e. issues to be covered) and on the style that your particular planning authority prefers. Several county and larger district planning authorities have published guidelines on environmental assessment and environmental statements for developments in their areas. Where these are available they should be followed. It may also be worthwhile using the local planning authority's archives to look at examples of applications and consents for similar types of developments. The guidance in the following paragraphs is inevitably of a general nature but should help you to ensure no important aspects are overlooked.

Description of the project

Describe the existing situation

- location and condition of site (what is there now and how is it affecting the local environment)
- site surroundings (focusing on any sensitive areas in the neighbourhood)
- present environmental quality (focusing on aspects which the project will affect)
- show the relationship of the project with any local Development Plans or Structure Plans.

Describe the project

- what is to be built
- what processes are to be installed, including types of materials to be handled, particularly any that are new to the site.
- any other changes to the site, e.g. new access, landscaping.

Alternatives considered

- an outline of alternatives studied and the main reasons for the proposed development, with particular reference to environmental effects.

Environmental effects

Environmental effects should be considered for all stages of the proposed project, including the temporary stages of site preparation and construction as well as the long term operation of the new activity.

Guidance on assessing the fate of releases in the environment is given in "Released Substances and their Dispersion in the Environment" (Ref. 1).

Air quality

Mindful of the distinctions between planning and IPC/IPPC, the information required by the planning authority about emissions to air is evidence that the project would not lead to any harmful levels of air pollution or nuisance. Detailed air dispersion modelling is unlikely to be required for relatively small-scale projects. However, if your project has substantial releases to air or there are sensitive sites in the neighbourhood, such as a hospital or a SSSI, more detailed predictions of air pollutant concentrations may be required.

- list all planned releases and outline how they will be controlled to ensure safe concentrations in the neighbourhood.
- put the emissions into context with other sources and indicate the project's contribution to local air pollution levels.

The effects of any deposition on the ground from airborne emissions, e.g. dusts, should be included here.

Water quality

- List any liquid effluents to be discharged directly to water courses or to sewer.
- Describe how they are to be controlled to avoid any harmful pollution.
- Put the discharges into context with any other local sources and indicate the project's contribution to local water quality.

Land-based waste disposal

List types of wastes and describe means proposed for any treatment, recovery, recycling or disposal, whether on-site or off-site.

Water supplies

Note any additional water consumption, compared with present site; whether there is to be any direct abstraction and the effect this would have on the surface water or aquifer concerned.

Energy supplies

List any additional energy supplies required, compared with present site, whether as direct fuels or grid electricity, and any knock on effects on local supply systems.

Soil and groundwater protection

Describe the means to be used to prevent contamination of soil and groundwater.

Visual appearance and lighting

Describe how the development will appear by day and by night; how it will affect the outlook for neighbouring sites, housing and other vantage points, including heights of new buildings, structures and chimneys, likelihood of visible steam plumes from chimneys or cooling towers.

Noise

Describe the measures that will be taken to avoid disturbance to the neighbourhood.

Traffic

Provide estimates of traffic expected during the construction and operational phases and describe measures that will be taken to avoid disturbance and congestion in the neighbourhood.

Ecology

Describe any likely effects on flora and/or fauna and measures to be taken to minimise such disturbance.

Hazards and risk assessment

Describe any significant hazards in the proposed project and the measures that will be taken to control the risks of harm to people or the environment.

(Advice on this aspect is not included in this guide. Useful references are given in Annex A.)

Economic and social effects

- provide estimates of temporary and permanent jobs created.
- describe any other gains to the local economy, e.g. material and services supplies
- describe any effects on public amenity, e.g. changes to rights of way or access or disturbance to cultural heritage sites, and what will be done to alleviate them.

4.5.2 Environmental reports for IPC/IPPC applications

The information required to support an IPC/IPPC application is primarily focused on releases from the process/installation and the means that will be used to control those releases. In this respect, the range of environmental effects to be described is more limited than the scope of the environmental report for a planning application, but more detailed information is required about the technical means that will be used to control such releases to acceptable limits. IPPC is also concerned with energy consumption, waste minimisation, measures to prevent accidents and site restoration.

Another distinction is that the planning process will judge the effects of environmental pollution primarily in terms of its acceptability at the proposed site. The IPC/IPPC authorising process will also consider whether the technical means for controlling releases, waste production and energy use are consistent with the current requirements of "best available techniques" for minimising harm to the environment as a whole. As a minimum, the control measures must ensure that the project would not lead to any breach of local environmental quality limits or compromise any national programmes for emission reductions.

Regulations and official guidance are available for IPC applications, but UK arrangements for implementing the IPPC Directive are still (December 1998) at the consultation stage. The guidance in the following paragraphs is based on current requirements for IPC and the EAC's interpretation of the IPPC Directive requirements.

Description of the proposed process/installation

A detailed description of the proposed process showing all the processing stages in flow diagram form and plant layout, with quantities of all materials into and out of the process. Energy use and water intake should be included.

Alternatives considered

Describe the alternatives you have considered for the proposed process. Use the outcome of the procedures from Chapter 4 to show why your proposed installation is the best practicable outcome for the environment as a whole.

Planned releases to air

List all chimneys and vents with emissions during normal operation. For each release point show location of release with height, volume, temperature; identify substances released, amounts and rates of releases with patterns of release over time.

List all emergency relief vents. Provide the same information as above, based on estimated causes and frequencies of operation of emergency relief systems.

Fugitive emissions: provide estimates of vapour escapes from glands and seals; describe any potential sources of particulate emissions from materials handling.

Planned releases to water

List all discharge points to site treatment, to sewer or to a surface water course. For each release point show location with volume and temperature; identification of substances released; amounts and rates of releases with patterns of release over time.

Planned wastes

Describe the type, quantity and source of all wastes that the process will generate for land-based disposal.

Techniques to minimise releases to air and their effects on the environment

Describe the technical means to be used to minimise and control releases, including design of abatement equipment, process controls and monitoring. Where appropriate, compare proposed means with Environment Agency guidance for the process category.

Techniques to minimise releases to water and their effects on the environment

Describe the technical means to be used to minimise and control releases, including design of abatement equipment, process controls and monitoring. Describe any treatment in on-site or off-site wastewater treatment plants before final discharge to a surface watercourse, estuary or sea. Where appropriate, compare proposed means with Environment Agency guidance for the process category.

Techniques to minimise wastes and their effects on the environment

Describe the technical means used to minimise wastes and the methods to be used for on-site or off-site waste treatment and final disposal, including any recycling or recovery outside the installation. Where appropriate, compare proposed means with Environment Agency guidance for the process category.

Energy use

List and quantify all the energy used by the proposed process - direct fuel, grid electricity - and explain how these represent the most efficient use of energy for this type of process.

Environmental effects of the proposed process - emissions, wastes and energy

Releases to air

For each of the significant releases to air, provide estimates of the dispersion in the atmosphere and resulting ground level concentrations. Results should be compared with the present air quality and with any relevant air quality limits or local improvement plans to show that they would not lead to any risk to human health or environmental impairment.

Releases to water

For each of the significant releases, and after any waste water treatment, estimate the resulting concentrations or effects in the final receiving water. Compare the results with any relevant water quality limits or local quality objectives to show that they would not lead to any breach of limits or harm to the aquatic environment.

Wastes for land-based disposal

Describe the final fate of all the wastes from the process. Provide evidence that all the proposed waste disposal routes are licensed to handle the types of wastes.

Energy use (IPPC only)

From the energy use data, estimate the carbon dioxide releases from producing that energy.

Noise (IPPC only)

Describe all the measures to be taken to prevent noise or vibration affecting areas outside the installation and measures for noise monitoring.

The above advice on energy and noise is preliminary and subject to change when the regulations and official guidance for UK Implementation of the IPPC Directive are available.

Hazards and risk assessment

Describe any significant hazards in the proposed project and the measures that will be taken to control the risks of harm to people or the environment.

Advice on this aspect is not included in this guide. Useful references are given in Annex A.

References

5. Environmental assessment - a guide to the procedures, 1989, Department of the Environment.

Annex A

USEFUL REFERENCES

The following list of references are the principal laws, regulations and official guidance documents that are likely to apply to new industrial developments. The references are grouped under the main areas of regulatory control that have been mentioned in this guide.

This list of references is for general guidance only and to inform project managers of the very broad span of environmental legislation and official guidance that may be relevant to their development plans. It is not a comprehensive list of every item of legislation that might apply to any project. Project managers should obtain advice from the regulating authorities and specialist legal advisers to ensure that they are aware of all their legal obligations.

Integrated Pollution Control (IPC)

1. Environmental Protection Act 1990, Part I (as amended by the Environment Act 1995).

2. Environmental Protection (Prescribed Processes and Substances) Regulations 1991, SI 1991/472 and subsequent amendments, under same title: SIs 1992/614, 1993/1749, 1993/2405, 1994/1271, 1994/1329, 1995/3247, 1998/767.

3. Environmental Protection (Applications, Appeals & Registers) regulations 1991, SI 1991/507 as amended by SI 1996/667.

4. Technical Guidance Notes:
   M1 - Sampling facilities for monitoring particulate emissions
   M2 - Monitoring emissions of pollutants at source
   D1 - Guidelines on stack heights for polluting emissions
   A1 - Guidance on flaring
   A2 - Pollution abatement technology for reducing solvent vapour emissions
   A3 - Pollution abatement technology for particulate and trace gas removal
   A4 - Effluent treatment techniques

5. Integrated Pollution Control - a practical guide (DETR).

6. Process Guidance Notes:
   The first series of guidance notes for IPC processes were produced by the former HM Inspectorate of Pollution as Chief Inspector's Guidance Notes with the prefix IPR. As these Notes come up for review they are being replaced by the Environment Agency's series of IPC Guidance Notes with the prefix S2.

   Series 2

S2 1.01 to S 2 1.12	Fuel products and combustion processes
S2 3.01 to S 2 3.04	Mineral industry sector
S2 4.01 to S 2 4.04	Chemical industry sector
S2 5.01 to S2 5.04	Waste Disposal and Recycling sector

   Series 1

IPR2/1 to IPR2/12	Metals production and processing
IPR6/1 to IPR6/9	Other industries

These notes provide guidance on techniques and emissions control limits which are likely to satisfy the requirement to use "best available techniques not entailing excessive cost" (BATNEEC) for the processes prescribed in the IPC Regulations (Ref. 2 above).

Integrated pollution prevention and control (IPPC)

7. EC Directive 96/61/EC "concerning integrated pollution prevention and control".

8. DETR consultation paper - "UK implementation of EC Directive 96/61 on integrated pollution prevention and control" - July 1997.

9. DETR consultation paper - second consultation paper on UK implementation of the IPPC Directive - January 1998.

Land use planning

10. Town & Country Planning Acts 1971 and 1990.
Town & Country Planning (Scotland) Act, 1997.

11. Planning (Hazardous Substances) Act 1990.
Planning (Hazardous Substances) (Scotland) Act, 1997.

12. Planning (Hazardous Substances) Regulations 1992, SI 1992/656.
Town & Country Planning (Hazardous Substances) (Scotland) Regulations 1993.

13. DoE circular 11/92 - Planning controls for hazardous substances.

14. DoE planning policy guidance note 23 (PPG23) - Planning and pollution control.
SODD Planning Advice Note 51 - Planning and environmental protection.

Environmental assessment

15. EC Directive 85/337/EC on the assessment of certain public and private projects on the environment.

16. EC Directive 97/11/EC replaces and extends the scope of 85/337.

17. The Town and Country Planning (Assessment of Environmental Effects) Regulations 1988, SI 1988/1199.
Environmental Assessment (Scotland) Regulations 1988, SI 1998/1221.

18. DoE circular 15/88 - Environmental assessment.

19. DoE - Environmental assessment - a guide to the procedures, 1989.

20. DoE - Evaluation of environmental information for planning projects - a good practice guide, 1994.

21. Environment Agency - Environmental assessment - scoping handbook for projects.

22. Kent County Council - Environmental assessment handbook.

Air pollution control

23. Part 1 of the Environment Protection Act 1990 (Ref.1 above) also covers Local Authority Air Pollution Control (LAAPC).

24. Secretary of State's Process Guidance Notes for LAAPC processes: PG series.

25. The Clean Air Act 1993.

26. Air Quality Standards Regulations 1989, SI 1989/317.

27. Air Quality Standards (Amendment) Regulations 1995, SI 1995/3146.

Water pollution control

28. Water Resources Act 1991 (and Environment Act 1995).
(for water abstraction and direct discharges to controlled waters)

29. Water Industry Act 1991 *(for effluent disposal to sewer)*.

30. Surface Waters (River Ecosystem) (Classification) Regulations 1994. SI 1994/1057.

31. Water Quality Objectives: procedures used by the NRA for the purposes of the Surface Waters (Classification) Regulations.

32. Trade Effluents (Prescribed Processes and Substances) Regulations 1989, SI 1989/1156.

33. Trade Effluents (Prescribed Processes and Substances) (Amendment) 1990, SI 1990/1629.

Waste disposal

34. Environment Protection Act 1990, Part II.

35. Controlled Waste Regulations 1992, SI 1992/588.

36. Waste Management Licensing Regulations 1994, SI 1994/1056 as amended by SI 1995/288, SI 1995/1950, SI 1996/1279.

37. The Special Waste Regulations 1996, SI 1996/972.

38. DoE circular 19/91 - Environmental Protection Act 1990 - the duty of care.

39. DoE circular 14/92 - the Controlled Waste Regulations 1992.

40. Waste management paper - Licensing of waste management facilities (third edition, 1994).

Control of accidental releases

41. EC Directive 85/501/EEC on the major accident hazards of certain industrial activities *(known as the Seveso directive; being replaced by ref. 41).*

42. EC Directive 96/82/EC on the control of major accident hazards involving dangerous substances *(known as the COMAH or Seveso II directive).*

43. Control of Industrial Major Accident Hazards Regulations 1984, SI 1984/1902, amended by SI 1988/1462, SI 1990/2325, SI 1994/118 *(known as the CIMAH regs.).*

General references

The following publications provide references and explanations of legislation pertaining to pollution control and environmental management. They include an updating service or are re-issued regularly.

44. Butterworth's Environmental Regulations.

45. Croner's Environmental Policy and Procedures.

46. Manual of Environmental Policy - the EC and Britain, by Nigel Haigh.

47. NSCA Pollution Handbook, 1998.

Annex B

CHECKLIST FOR ENVIRONMENTAL REVIEW

The following list of questions and topics may be used as a prompt for preparing a checklist for the preliminary environmental review and for identifying the environmental aspects that you should consider when comparing process and site options.

NB No single checklist can cover all the environmental aspects of all types of projects or all locations. A specific checklist should always be prepared for each project.

PROCESS CONSIDERATIONS

1. **Substances**
   - 1.1 Will the process use any substances which are, or are likely to be, restricted? Consider raw materials, process chemicals and catalysts, packaging materials, cleaning materials.
   - 1.2 Do any of the substances to be used involve the use of restricted materials in their production? Could this limit their availability in the future?
   - 1.3 Will the process produce any substances which are, or are likely to be, restricted? Consider finished products, by-products, waste products.

2. **Energy**
   - 2.1 Is the process a major user of energy?
   - 2.2 How will energy be provided? e.g. national grid, on-site boilers, fired heaters, CHP, etc?

3. **Fresh water consumption**
   - 3.1 Will the process use substantial quantities of fresh water? Is an abstraction licence needed?

4. **Emissions to air**
   - 4.1 Will there be planned releases to air?
   - 4.2 What substances will be released?

5. **Liquid effluents**
   - 5.1 Will there be planned releases directly to surface waters?
   - 5.2 Will there be planned releases to sewer?
   - 5.3 What substances will be released?

6. **Wastes to land-based disposal**
   - 6.1 Will there be planned wastes for off-site disposal?
   - 6.2 What will be the nature of the wastes and the substances in them?
   - 6.3 What waste disposal methods will be used?
   - 6.4 Could the waste disposal methods lead to any risk of soil and groundwater contamination?

7. **Hazards**
   - 7.1 Will the development have inventories of hazardous materials that would qualify as a Major Hazard installation?
   - 7.2 Are there any other hazardous materials or processes that are potential risks to the environment?

8. **Noise and vibration**
   - 8.1 Are any aspects of the development likely to be noisy?
   - 8.2 What are the expected hours of operation?

9. **Lighting**
   - 9.1 Will there be illuminated outside areas?

10. **Visual appearance**
    - 10.1 What is the likely appearance of the development? E.g. area, heights, buildings or exposed plant, yards, open storage, etc.

11. **Transport**

11.1 What traffic will the project generate? Types and numbers of vehicles for movement of materials and people; traffic routes and properties affected.

11.2 What are the likely hours of work?

SITE CONSIDERATIONS

12. **Condition of site**

12.1 What is the history of the site?

12.2 Is there any evidence, from history or the current situation, that site may be contaminated?

12.3 Is there any evidence, from history or the current situation of the surrounding area, that could lead to contamination of the site?

12.4 What is the state of any site infrastructure that may become part of the project? e.g. drains, culverts, roadways, utility supply and distribution systems, fire protection systems, firewater run-off control, storage area bunding, waste water treatment systems.

13. **Site surroundings**

13.1 Would the development be consistent with the local development plan?

13.2 How close are houses?

13.3 How are they likely to be affected by the development? Noise, traffic, visual impact, smell, lighting, etc.?

13.4 Are there likely to be local concerns about hazardous substances or operations from the point of view of public safety or public health?

13.5 Are there are any neighbouring industrial or commercial activities that are likely to be affected by the project?

13.6 Are there any neighbouring horticultural or agricultural activities that are likely to be affected by the project?

13.7 Are there any sensitive sites in the vicinity that could be affected by the project? e.g. nature reserves, SSSIs, ancient monuments, parks and recreation areas.

13.8 Are there any sensitive activities in the vicinity that could be affected by the project? e.g. hospitals, schools.

14. **Air quality**

14.1 What is the local air quality?

14.2 Is the project likely to have a significant effect on local air quality?

14.3 What are the local plans for air quality improvement?

15. **Water quality**

15.1 What is the quality of any watercourse that would receive effluent from the project, either directly or via a local sewage treatment works?

15.2 What are the local plans for improving the quality of local watercourses?

15.3 Does the local sewage treatment works have capacity for treating effluent from the project?

16. **Waste disposal facilities**

16.1 What facilities are available for industrial waste disposal? e.g. landfill, incineration, recovery of recyclable materials.

LEGAL CONSIDERATIONS

17. **Legislation**

17.1 What legislation applies to the environmental aspects of the project? Planning - IPC/IPPC - Water Resources Act - etc.

18. **Relevant authorities**

18.1 Identify the local officers responsible for enforcement of the legislation identified under 17.1 above.

ECONOMIC & SOCIAL CONSIDERATIONS

19. **Employment**

19.1 What is the number and type of any new temporary and permanent jobs that will be created by the project?

19.2 What number of new jobs are likely to be available to local residents?

19.3 What other economic benefits are likely in the local community? E.g. raw material and other supplies, contract services.

20. Local interests

20.1 Are there any issues of local concern that would be aggravated by the project, e.g. traffic problems, local air or water pollution, existing industrial activities with a poor environmental record?

Annex C

CONCEPT RANKING

This is an alternative approach to the "forced ranked pairs" technique described in section 4.3 for comparing and ranking the environmental effects of two or more project options.

The method is as follows:

1. Identify 4 or 5 main *categories* of environmental impacts which show differences in effects between the options, e.g. - pollution; - noise; - visual impact; - transport.

2. Assign an *importance weighting* to each category by allocating 100 units between the categories. This is a value judgement reached by consensus or averaging among the views of the project team members and advisers, e.g: - pollution (40); - noise (15); - visual impact (20); - transport (25).

3. Within each category, identify a number of *impact parameters* which show differences between the options. Allocate the importance weighting of the category among the parameters, e.g. in the category "pollution": - air quality (25); - river water quality (15), i.e. a total of 40.

The values allocated to each *impact parameter*, as a result of this procedure, reflect the project team's view of the relative importance of each *impact parameter* in the overall context of the project.

4. For each option, assign a *significance factor* to each impact parameter on a scale from 1 (negligible) through 2 (minor significance), 5 (moderately significant), 8 (very significant) to 10 (extremely significant).

These *significance factors* are the team's judgement of the significance of each impact parameter for the option under consideration. E.g. the impact parameter "visual impact - views from houses" has the same *importance factor* for all options but may be of *minor significance* for one site option and *very significant* for another.

5. Record the significance factor for each option and impact parameter in a table, as shown in the example below.

6. Compile *impact scores* by multiplying the *importance weighting* by the *significance factor* for each parameter and option.

7. Sum the impact scores for all parameters for each option to give a *total impact score* for each option. The *total impact scores* are then an indication of the *relative ranking* of the overall environmental impact of the options. This is illustrated in the table.

Example of Concept Ranking Calculation

The stage of allocating 100 units of "importance weightings" to the impact categories and parameters is likely to produce a similar ranking order of environmental impacts to the "forced ranked pairs" method described in section 4.3. The concept ranking technique goes a step farther by combining "importance weighting" with "significance factor" to produce an "impact score".

In this example option 2 has the lower total of impact scores and, on that basis, is the preferred option. The difference in total impact score is some 30% which is a fairly clear margin. If the difference were less, say below 15%, it would be appropriate to look at the individual impact scores for the higher weighted impacts and test the sensitivity of the total impact scores to small changes in the significance factors.

Whatever method you use to decide which of two or more options have the lesser overall environmental impact, all the information you have used should be displayed to justify your decision. All your reasoning should be apparent to others. All procedures involving the use of numbers to combine value judgements carry the risk that the numerical answers will give a false sense of precision and mask all the uncertainties that are inherent in environmental assessment.

Impacts	Importance Weighting	Option 1		Option 2	
		Overall Significance Factor	Impact Score	Overall SIgnificance Factor	Impact Score
Category 1 - Pollution	**40**				
Impact Parameter 1 - air quality Impact Parameter 2 - river water quality	25 15	8 5	25 x 8 = 200 15 x 5 = 75	8 2	25 x 8 = 200 15 x 2 = 30
Category 2 - noise	**15**	**5**	**15 x 5 = 65**	**5**	**15 x 5 = 65**
Category 3 - visual impact Impact Parameter 1 - views from house Impact Parameter 2 - views from recreational areas **Category 4 - transport** Impact Parameter 1 - congestion Impact Parameter 2 - disturbance	**25** 20 5 **25** 15 10	 8 5 8 5	 20 x 8 = 160 5 x 5 = 25 15 x 8 = 120 10 x 5 = 50	 5 2 5 2	 20 x 5 = 100 5 x 2 = 10 15 x 5 = 75 10 x 2 = 20
Total Impact Score			695		500